The SKYWARN® Spotter and the Spotter's Role

The United States is the most severe weather-prone country in the world. Each year, people in this country cope with an average of 10,000 thunderstorms, 5,000 floods, 1,200 tornadoes, and two landfalling hurricanes. Approximately 90% of all presidentially declared disasters are weather-related, causing around 500 deaths each year and nearly $14 billion in damage.

SKYWARN® is a National Weather Service (NWS) program developed in the 1960s that consists of trained weather spotters who provide reports of severe and hazardous weather to help meteorologists make life-saving warning decisions. Spotters are concerned citizens, amateur radio operators, truck drivers, mariners, airplane pilots, emergency management personnel, and public safety officials who volunteer their time and energy to report on hazardous weather impacting their community.

Although, NWS has access to data from Doppler radar, satellite, and surface weather stations, technology cannot detect every instance of hazardous weather. Spotters help fill in the gaps by reporting hail, wind damage, flooding, heavy snow, tornadoes and waterspouts. Radar is an excellent tool, but it is just that: one tool among many that NWS uses. We need spotters to report how storms and other hydrometeorological phenomena are impacting their area.

SKYWARN® spotter reports provide vital "ground truth" to the NWS. They act as our eyes and ears in the field. Spotter reports help our meteorologists issue timely, accurate, and detailed warnings by confirming hazardous weather detected by NWS radar. Spotters also provide critical verification information that helps improve future warning services. SKYWARN® Spotters serve their local communities by acting as a vital source of information when dangerous storms approach. Without spotters, NWS would be less able to fulfill its mission of protecting life and property.

Spotter Reporting Procedures

Effective spotter reports are a critical component of NWS severe weather operations. NWS meteorologists use science, technology, training, experience, and spotter reports when making warning decisions. An effective spotter report is one that is timely, accurate, and detailed. Spotters should use the following guidelines when reporting:

- Follow the specific reporting guidelines for your area.
- Remain calm, speak clearly, and do not exaggerate the facts.
- If you are unsure of what you are seeing, make your report, but also express your uncertainty.
- Your report should contain the following information:
  - WHO you are: trained spotter
  - WHAT you have witnessed: the specific weather event
  - WHEN the event occurred: NOT when you make your report
  - WHERE the event occurred, (not necessarily your location) using well known roads or landmarks

Immediate, real-time reports, are most helpful for warning operations, but delayed reports are also important, even days after an event. Delayed reports are used for climatological and verification purposes.

Weather events should be reported according to the instructions provided by your local NWS office. Here are some general guidelines on what to report.

Tornadoes

- What damage did you observe?

- How long was it on the ground? When did it start and end?

- How wide was it? How far did it travel if known?

Flash Flooding

- Report flooded roadways, rivers and streams, giving approximate water depth.

- Does the flooding consist of standing water or is it flowing?

- Is the water level continuing to rise, staying steady or falling?

- Is the flooding occurring in a known flood prone area?

- Any damage from the flooding or mud slides?

Wall Clouds

- Report if clouds are rotating and how long they have existed.

Funnel Clouds

- Watch for organization, persistence and rotation.

Lightning

- Only report lightning when damage or injuries occur.

Winter Weather

- Report any occurrence of freezing rain, ice accumulation and damage.

- How much heavy snow accumulation is there and is there any damage?

- Do blizzard conditions exist: winds 35 mph or more AND visibility ¼ mile or less?

Wind

- Report estimated or measured wind speed and wind damage.

- Wind speed estimation is difficult. A detailed description of moving objects or damage is often more useful.

- Details to submit for tree damage:

  - What is the height and diameter of the branch, limb or tree that was broken or blown down?

  - Was the tree healthy or decayed?

  - What type of tree was damaged, e.g., hardwood or softwood?

- Details to submit for damage to structures.

  - Is the damage to a well-built structure or a weak outbuilding?

  - What is the main building material for the structure: wood, brick, metal, concrete, etc.?

  - If the structure is a mobile home, was it anchored down?

Hail

- Report the size of the largest stone and any damage.

  - To estimate size, compare hail to well known objects such as coins or balls, but not to marbles, or measure the hail with a ruler.

Marine Hazards

Report the following marine events:

- Waterspouts: you must observe rotation

- Squall lines

- Heavy freezing spray

- Wave heights and winds that differ significantly from forecasted conditions

- Hydrometeorological phenomena that are not in the current marine forecast, e.g., thunderstorms, dense fog

- Waves greater than twice the size of surrounding waves

- Tsunami inundation and any damage

- **Coastal Flooding:** Inundation of people, buildings, and coastal structures on land at locations that under normal conditions are above the level of high tide

- **Lakeshore Flooding:** Inundation of land areas along the Great Lakes over and above normal lake levels

- **High Surf:** Large waves breaking in the surf zone with sufficient energy to erode beaches, move large logs, wash over jetties or exposed rocks, etc.

Other Environmental Hazards

- Dense fog: visibility ¼ mile or less

- Dust storms: visibility ¼ mile or less

- Volcanic ash accumulation and any damage

- Any injuries or fatalities as a direct result of weather

Spotter Safety Tips

The environment in and around severe storms is a dangerous place. Even though tornadoes are an obvious danger, other life-threatening thunderstorm hazards, such as lightning and flash floods can be just as deadly. Spotter reports are vital to your community and the NWS, but your safety should be your number one priority!

Before venturing out, you need to be aware of the hazards of thunderstorms and the recommended practices to minimize risk. As a spotter, it is your responsibility to stay safe while spotting. Please following the guidelines below for your personal safety and for the safety of those around you.

- Personal safety is the primary objective of every spotter.
- Adhere to the concept of ACES, defined below, at all times.
- Obey federal, state, and local laws and directives from public safety officials.
- Never put yourself in harm's way. This includes attempting to walk or drive over obstructions such as flooded roadways and downed power lines, and positioning yourself under objects that have a potential to fall or be blown over due to severe weather.

ACES stands for **Awareness, Communication, Escape Routes, and Safe Zones. ACES** is a concept commonly used by emergency management personnel. If you remember ACES, you can remain safe in any situation, including spotting.

Bolivar Peninsula, TX, September 20, 2008: Damaged houses, debris and downed power lines resulting from Hurricane Ike. Photo by Jocelyn Augustino, FEMA.

Awareness means you are constantly observing the situation around you. This type of observation is sometimes referred to as situational awareness. Continuously monitoring the risks around you can save your life, especially in rapidly changing weather conditions. Knowing that there is a river crossing, or observing the street is lined with power poles and trees, can prepare you for the hazards of severe weather. When you are aware of the imminent threats, and you are thinking ahead about possible outcomes, you can position yourself better to minimize these threats.

Communicating your whereabouts to others on a regular basis and having multiple lines of communication available can keep you and others safe from hazards.

Escape Routes are vital when you are entering a potentially dangerous area. As part of awareness, note the escape routes available to you, making sure you always have more than one and the safest way to get to that escape route.

If the event you cannot get to escape routes due to rapidly changing conditions, find your closest safe zones or shelters. Safe zones are the areas where you will be safest if you need to get to immediate shelter. Knowing these locations will limit your risk.

Remembering ACES: to remain aware of your surroundings, have open lines of communication, know your escape routes, and know your safe zones wherever you are can increase your safety.

Here are a few basic tips that could save your life if you are watching a storm from your vehicle.

- Keep a buffer zone between you and the storm to allow for changes in storm movement and to keep your options open for an escape route.

- Travel in pairs so the driver can concentrate on driving, and so you can observe multiple areas of the storm.

- Always know where you are in relation to the storm, and which way the storm is moving. Remember that storms can change direction and speed.

- Never drive through the core of the storm, e.g., through heavy rain and/or hail, to get a better vantage point.

- Have a source of current local weather information, such as a NOAA Weather Radio, to be sure you have critical storm information.

- If possible, observe a storm from a four-way intersection to facilitate escape in multiple directions.

- Nighttime storm spotting in a vehicle can be dangerous and is not recommended. It is difficult to observe key storm features at night and harder to safely maneuver around the storm.

- Be alert for emergency vehicles, pedestrians, and other road and traffic hazards.

- If you stop, remember to watch for traffic and to be aware of the potential effects of the storm. Pull out of traffic and keep away from trees, power lines, and signs.

- Never stay in a vehicle under large trees or signs in high wind conditions. Get into a sturdy structure.

- Always be prepared for all of the hazards associated with thunderstorms.

- If close to or in the path of a storm, do not turn off your vehicle.

Tornado Safety

Tornadoes are violently rotating columns of air attached to a thunderstorm and in contact with the ground, whether or not a condensation funnel is visible to the ground. Debris or dirt swirling on the ground, under an area of cloud base rotation, may be a clue that it is a tornado and not a funnel cloud or gustnado. The high winds and flying debris associated with a tornado pose a significant threat to a spotter. Here are a few safety tips if you encounter a tornado:

- Watch for other tornadoes that could form in the vicinity of the tornado you are watching.

- Never try to outrun a tornado in an urban or congested area. Immediately get into a sturdy structure after parking your car out of the traffic flow.

- Do not take shelter under bridges or overpasses. These structures do not offer protection and could increase the chance of injury or death.

- If you are caught outdoors, seek shelter in a basement, shelter or sturdy building. If you cannot quickly get to a shelter, immediately get into a vehicle, buckle your seat belt and try to drive to the closest sturdy shelter. If flying debris occurs while you are driving, pull over and park. You have the following options as a last resort:

  - Stay in the car with the seat belt on. Put your head down below the windows, covering with your hands or a blanket if possible.

  - If you can safely get noticeably lower than the level of the roadway, exit your car, and lie in that area, covering your head with your hands.

- Flying and falling debris is the biggest hazard in a tornado. To be safe, you should get inside, get down and cover up. Underground or in a Safe Room is your first choice. If no underground shelter is available, get to the center of a sturdy building on the lowest level. Put as many walls between you and the tornado as possible. Stay away from windows and doors. Cover up to help minimize being injured by flying or falling debris.

Flash Flood Safety

Turn Around Don't Drown! Thunderstorms can produce torrential rain over a short period resulting in flash flooding. Flooding is particularly dangerous at night when it is harder to see the road is flooded and even harder to tell how deep the water is. Flooding causes more fatalities each year than any other thunderstorm hazard. More than half of all flood-related drownings occur when a vehicle is driven into hazardous flood water. Use these safety tips and facts to avoid being a victim of a flash flood:

- **Turn Around Don't Drown!** Do not attempt to drive or walk across a flooded road or low water crossing. You cannot be sure about the depth of the water or the condition of the roadway. The road might be washed out.

- Two feet of moving water will carry away most vehicles.

- Six inches of fast-moving water can knock you off your feet.

- If your vehicle is suddenly caught in rising water, leave it immediately and get to higher ground.

- Be especially vigilant at night when flash floods are harder to recognize.

Street flooding in downtown Reno, NV, January 2006. Photo from NOAA.

Lightning Safety

When Thunder Roars, Go Indoors! Lightning is an underrated killer. Nearly as many people lose their lives to lightning strikes as they do to tornadoes, but because lightning typically hits just one or two people at a time, fatalities due to lightning receive less publicity. Lightning occurs with every thunderstorm and is the most common weather hazard facing spotters. As a spotter, you are frequently positioned in the open or on a hill top, making you especially vulnerable to lightning. Here are a few important safety guidelines for dealing with lightning:

- Remain in a hard-topped vehicle or an indoor location for at least 30 minutes after you hear the last thunder clap. If you use radio equipment, avoid contact with it or other metal inside your vehicle to minimize the impacts should lightning strike.

- If you are out on the water and skies are threatening, get back to land and find a fully enclosed building or hard-topped vehicle. Boats with cabins offer a safer but not perfect environment. You are safer if the boat has a properly installed lightning protection system. If you are inside the cabin, stay away from metal and all electrical components.

- Do not use a corded phone during a thunderstorm. Use a cordless phone or cell phone for all calls.

- Lightning victims do not carry an electrical charge, are safe to touch, and need urgent medical attention. If a person has stopped breathing, call 9-1-1 or your local emergency phone number and begin CPR if the victim is not breathing.

Downburst Wind Safety

Damaging thunderstorm straight line winds known as downbursts are another hazard facing a spotter. A downburst is a strong downdraft with an outrush of damaging winds on or near the ground. Most of the wind damage done by severe thunderstorms is caused by downbursts. Downburst winds may exceed 100 mph in the most intense storms, and may cause damage similar to a tornado. Here are some tips to stay safe in and around downbursts:

- Keep a firm grip on your vehicle's steering wheel to maintain control. Downbursts can occur suddenly with an abrupt change in wind speed and direction.

- If you can do so safely, point your vehicle into the wind to minimize the risk of the vehicle being blown over.

- Be prepared for sudden reductions of visibility due to blowing dust or heavy rain associated with downbursts.

- Point spotters observing from a substantial building should move away from windows as the downburst approaches.

Hail Safety

Large hail can cause serious injuries and damage to vehicles and buildings. Although fatalities attributed to hail are rare, it is the costliest weather element in the United States with an average of more than a billion dollars in agriculture and property damage each year. The costliest U.S. hailstorm caused around $2 billion dollars in damage in the St. Louis Metropolitan Area on April 10, 2001. Below are some points about hail that could minimize damage to your vehicle when storm spotting.

- Substantial structures and buildings such as a garage offer the best protection from hail.

- If in a vehicle, avoid those parts of the storm where large hail is occurring.

- Hard-top vehicles offer good protection from hail up to about golf ball size. Larger hail stones will damage windshields.

A photo—or video—is worth a thousand words!

Spotters often observe amazing weather phenomena. Whether it is a tornado in the distance or ice taking down power lines, your local NWS Forecast Office learns a great deal from spotter photos and videos. If you are willing to allow the NWS to use your photo or video in our education and outreach efforts, please state that NWS has permission to reprint when you submit the file to us. If you would like us to credit you on the image, we will gladly do that. Multimedia files are a tremendous resource when conducting our spotter training courses and our weather safety education efforts.

Send NWS Your Home Weather Station Data Service

The Citizen Weather Observer Program (CWOP) is a private-public partnership that allows people with computerized weather stations and always on Internet access to send their weather information to a special data server that collects weather observations from around the country. Your data can then be used by computer forecast models to produce short term forecasts (3 to 12 hours into the future) of weather conditions in your region.

Visit http://wxqa.com to register for an ID and learn which weather stations and software packages work with the CWOP network.

Check out http://www.met.utah.edu/mesowest from the University of Utah to see a plot of surface observations available from home weather stations, the National Weather Service, state Departments of Transportation, the Federal Aviation Administration, and the U.S. Forest Service.

CoCoRaHS: Community Collaborative Rain, Hail and Snow Network

CoCoRaHS is a unique non-profit, community-based network of volunteers of all ages and backgrounds working together to measure and map precipitation: rain, hail and snow. By using low-cost measurement tools, stressing training and education, and accessing an interactive web-site, volunteers provide the highest quality data for natural resource, education and research applications. The program is open to anyone who has an interest in weather and would like to share their precipitation data with others. Requirements of the program include:

- A 4 inch diameter, high-capacity rain gauge purchased through the CoCoRaHS website or from other sources

- A ruler to measure hail size: stone diameter

- Hail pads (not used in all states)

- Internet access; some states have a phone number where volunteers without computers can leave a message with their precipitation information. The data is later entered online.

To join the CoCoRaHS program, go to: http://www.cocorahs.org. The precipitation data is available for anyone to see online.

Thunderstorm Basics

Part of the fascination many people have with thunderstorms is the mystery that surrounds them. Leading researchers are still learning about many of the phenomena associated with thunderstorms. In order to understand thunderstorm-related spectacles like tornadoes, lightning, and hail you must have a basic knowledge of thunderstorm characteristics.

Thunderstorm Climatology

At any given moment, there are thousands of thunderstorms occurring worldwide. Most of these storms are beneficial, bringing needed rainfall. A small percentage of the storms become severe, producing large hail—1 inch in diameter or larger—strong wind gusts of 58 mph or greater, or tornadoes.

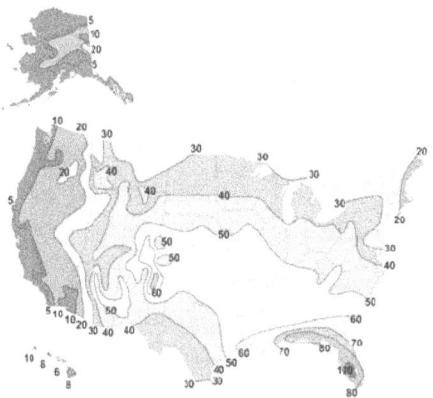

Average annual thunderstorm days per year

Although the area stretching from Texas to Minnesota has the greatest known occurrences of U.S. severe thunderstorms, no place in the United States is immune to the threat of severe weather.

Thunderstorm Ingredients

All thunderstorms require the following three ingredients: moisture, instability, and lift. Organized severe thunderstorm events also require vertical wind shear.

Moisture forms the clouds and precipitation in thunderstorms. Primary moisture sources include the Atlantic and Pacific Oceans and the Gulf of Mexico. The Great Lakes also can provide moisture for thunderstorms. In the Midwest, evaporation from farmlands can enhance low-level moisture.

Atmospheric stability is a measure of the atmosphere's tendency to enhance or deter vertical motion. In unstable conditions, a lifted parcel of air will be warmer than the surrounding air at that altitude. Because it is warmer, it is less dense and can rise more. Thus instability favors a storm's updrafts and downdrafts.

Lift provides the mechanism for the air to rise, starting the thunderstorm process. Sources of lift include cold fronts, warm fronts, drylines, thunderstorm outflow boundaries, and flow up the slopes of topography.

Vertical wind shear is the change in wind speed and direction with height. This effect is typically strongest near the surface, though it can be very strong at higher levels in the atmosphere near upper level jets and fronts. Generally, the greater the instability, the stronger the updrafts and downdrafts may become. The greater the vertical wind shear, the better the chance of storms becoming organized and long-lived. Vertical wind shear through a deep layer (3-5 miles above ground) also can induce rotation—at times intense—in the storm's mid levels. Storms developing in weak-shear environments still can produce brief hail and microbursts, and even weak tornadoes.

Low-level vertical wind shear (in the lowest 1 mile or less of the atmosphere) can help to generate low-level rotation in a storm. Tornadoes and severe winds are most often related to the strength of the low-level rotation. When the low-level shear increases in strength, the strength of the low-level rotation increases, along with the likelihood of tornadoes.

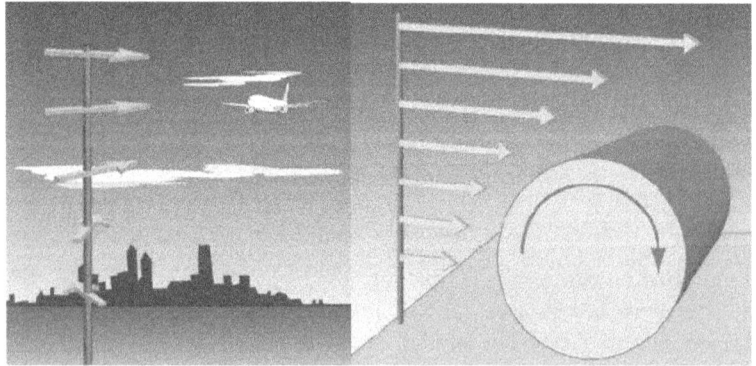

Examples of directional shear (left) and speed shear (right).

Thunderstorm Life Cycle

Thunderstorms generally last 30-60 minutes, but because they continue to form new updrafts, they can last for over 8 hours. Thunderstorms have three distinct stages:

1. Developing Stage (Cumulus or Towering Cumulus)
   - Updraft, upward moving column of air, develops.
   - Storm begins to produce precipitation within the upper portion of the cloud.

2. Mature Stage
   - Updraft and downdraft coexist.
   - Downdraft reaches the ground as a spreading of rain-cooled air called the cold pool. The leading edge of the cold pool is called the gust front.
   - Top of updraft forms an anvil-shaped cloud as air spreads outward.

3. Dissipation Stage
   - Downdraft is dominating.
   - Loses favorable inflow as the gust front moves out a long distance from the storm.
   - Sometimes shows an "orphaned anvil," the remnants of an anvil with the storm dissipated below.

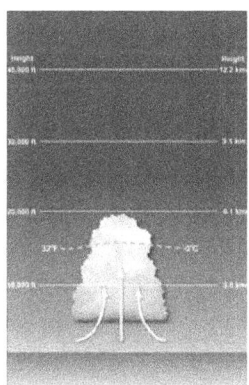 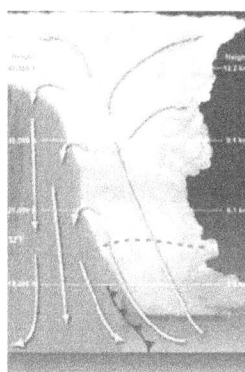 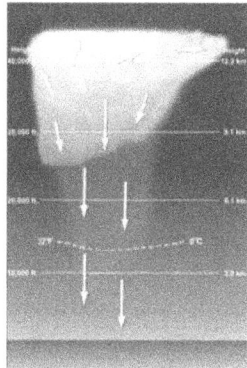

Towering Cumulus Stage   *Mature Stage*   *Dissipating Stage*

Severe Thunderstorms

A severe storm has at least one of the following:

- Hail that is 1 inch or larger: the size of a quarter
- Wind gusts of at least 58 mph or higher: 50 knots
- Tornado

The Storm Prediction Center issues Tornado and Severe Thunderstorm Watches. Local weather forecast office issue Tornado or Severe Thunderstorm Warnings.

Watch: Conditions are favorable in a region for severe thunderstorms or tornadoes to develop or to move into your area. Watches are generally issued for a 4-8 hour period. Be ready for storms in the near future. Take action to protect property before the storms arrive in your area.

Warning: There is either an imminent threat or an actual occurrence of large hail, damaging winds or a tornado. When a warning is issued, respond immediately to protect life and property. Warnings are generally issued for 30-60 minutes.

Thunderstorm schematic

Features Indicating Strong/Severe Storms

Anvil: The anvil is the elongated cloud at the top of the storm that spreads downwind with upper level steering winds. The anvil will appear solid, not wispy, and will have sharp, well defined edges.

Overshooting Top: The overshooting top is the dome of cloud directly above the main storm updraft tower and the anvil. If the overshooting top is persistent and lasts for 10 minutes or longer, it is generally a sign of a very strong thunderstorm updraft.

Main Storm Tower: The "trunk" of the storm is the visible updraft of the storm from its base near the ground to just below the anvil. This part of the storm can show:

- Vertically oriented tower, with sharp, well defined edges

- Solid, cauliflower appearance

- Visible rotation of the middle and lower levels, and possibly striations evident in the clouds

Rain-free Base: The area below the main storm tower. It is generally on the south or southwestern flank of a storm.

Wall Cloud: A wall cloud is an isolated lower cloud attached to the rain-free base and below the main storm tower. Wall clouds often are on the trailing side of a storm. For example, with a storm that is moving to the north or northeast, the wall cloud typically is on the south or southwest side of the storm. With some storms, the wall cloud area may be obscured by precipitation. Wall clouds associated with potentially severe storms:

- Usually persist for 10 minutes or more

- Often, but not always, rotate visibly

- Sometimes are accompanied by obvious rising or sinking motion of cloud piece

Shelf Cloud: A shelf cloud is a low, horizontal, banded cloud attached to the base of the parent cloud, usually a thunderstorm. Rising cloud motion often can be seen in the leading, outer part of the shelf cloud, while the underside often appears turbulent and wind-torn. Generally, a shelf cloud appears on the leading edge of a storm.

Roll Cloud: A roll cloud is low, horizontal, tube-shaped and relatively rare. It differs from a shelf cloud by being completely detached from other cloud features.

Severe/Strong Storm Features Checklist

Upper-level storm features visible at long distances from the storm:

- A solid-looking overshooting top persisting for 10 minutes or more

- Overshooting tops that may dissipate, followed by new ones

- A solid-looking anvil with sharply defined edges

Why? An overshooting top is a signal of a strong updraft. If the top persists for at least 10 minutes, it is a sign that the storm is continuing to strengthen. If the overshooting top suddenly collapses, a burst of precipitation, hail, or damaging wind may be imminent. As the storm weakens, it will take on a wispier, fuzzy appearance.

Why? A solid-looking storm tower indicates a strong updraft along with a favorable shear environment. A flanking line indicates the storm is drawing air from many miles away and likely will sustain itself or intensify for some time.

Low-level storm features can be seen when you are close to a storm:

- Rain-free cloud base with a large and solid-looking storm tower above

- Wall cloud persisting for 10 minutes or longer, especially if it is obviously rotating

- Rapid vertical motion (up or down) within the wall cloud or other areas of the rain-free cloud base

Why? A rain-free cloud base indicates a strong updraft, where precipitation and hail is not heavy enough to fall to the ground. When a rotating wall cloud is present, there is a much higher potential for tornado development. Wall clouds begin to rotate as the larger scale circulation, or mesocyclone, 2-10 miles in diameter, develops toward the surface. When combined with the proper atmospheric conditions, this pattern supports tornadoes.

Thunderstorm Types

Thunderstorms can be categorized by their physical characteristics: the presence or absence of rotation, the number of location of updrafts and downdrafts present.

There is a continuous spectrum of storms in the sky. At times, it is difficult to place a storm into a specific category. A storm may move from one category to another. These five types are often useful in describing storms:

- Ordinary or Single Cell Storm: Single cell storms are short lived, and usually not severe.

- Pulse Storm: A Pulse Storm is a single-cell thunderstorm that is usually not strong; when it is of substantial intensity, it produces severe weather for short periods of time. Such a storm weakens and then generates another short burst or pulse.

- Multicellular Cluster: This type is the most common storm, consisting of a group of ordinary cells at various stages of the thunderstorm life cycle.

- Multicellular Line: This category is a long line of storms with a continuous, well developed gust front along the leading edge.

- Supercell: A supercell is a highly organized thunderstorm with an extremely strong updraft. They exhibit persistent storm-scale rotation of the updraft-downdraft couplet or mesocyclone.

Ordinary or Single Cell Storm

This storm forms when there is weak shear in the atmosphere. Characteristics of this kind of storm include:

- Short life, generally 30-45 minutes
- Downdraft that forms within 15-20 minutes after cell initiation
- Updraft that weakens in 25-30 minutes, outflow stabilizes
- Small hail, usually not severe
- Gusty winds, usually not severe

The Pulse Storm

- Short life, generally 30-45 minutes

- Usually not severe, but given the right environmental conditions, these storms can create:

  - Brief, small to moderate size hail

  - Downburst winds, usually less than 70 mph

  - A weak tornado

- Damage is isolated

The Multicell Cluster Storm

- Most common type of thunderstorm

- May last for several hours

- Consists of a group of cells moving as a single unit

- Contains cells in different stage of the thunderstorm life cycle

- Occasionally may contain supercells

The new updrafts in the cluster form in an area of persistent lifting where air converges in low levels, such as:

- Cold or warm front

- Dry line

- Outflow boundary from nearby storms

- Higher terrain features than the surroundings

Typically, cells will develop in the lifting zone and move with the mid and upper level winds as it matures and dissipates, with new cells continuing to develop. In one typical scenario during the spring and summer months:

- New cells initiate on west or southwest edge of cluster

- Dissipating cells weaken on east or northeast edge of cluster

- Each cell lasts 20-30 minutes

- Clusters as a whole often last an hour or more

Given the right conditions, the cells can become severe within the multicellular cluster producing:

- Brief, small to moderate size hail

- Downburst winds

- Weak tornadoes

- Heavy rainfall in a short time

The Multicell Line Storm

- Frequently called squall line

- A long line of storms with individual storm outflows merging to produce a continuous, well developed gust front marking the leading edge of rain-cooled air

- Line of storms often oriented north-south or northeast-southwest and usually move toward the northeast, east or southeast

- May be embedded along the line

Multicellular Line

Individual thunderstorm updrafts and downdrafts along the line can become severe, resulting in large hail and episodes of damaging outflow winds that move rapidly ahead of the system.

Given the right environmental conditions, multicell line storms can produce:

- Strong downburst winds

- Heavy rainfall

- Moderate-sized hail

- Occasional tornadoes

The Supercell

Classifying supercellular storms is subjective. There is a lot of research being done in this area. We do know the following about supercells:

- Highly organized storm with rotation inside

- Updrafts can attain speeds more than 100 mph

- Can produce extremely large hail and strong, violent tornadoes

- Rear-flank downdraft can produce damaging outflow winds in excess of 100 mph

It is essential to become familiar with the visual aspects of these intense thunderstorms. Two important characteristics that distinguish supercells from ordinary thunderstorms are:

- Persistent rotation at the rain-free base

- A rear-flank downdraft (RFD): region of dry air wrapping along the back portion of the circulation within the storm

The presence of a strong rotation in the storm greatly enhances updraft intensity, persistence and overall storm organization. The temperature and moisture characteristics of the RFD and its evolution during a storm's lifetime have been shown to play a crucial role in tornado formation. Supercells can produce the following elements:

- Large hail and potentially torrential rainfall immediately adjacent to the storm updraft

- Smaller hail and lighter rainfall at greater distances from the updraft

- RFDs producing strong, sometimes damaging outflow

- Tornadoes

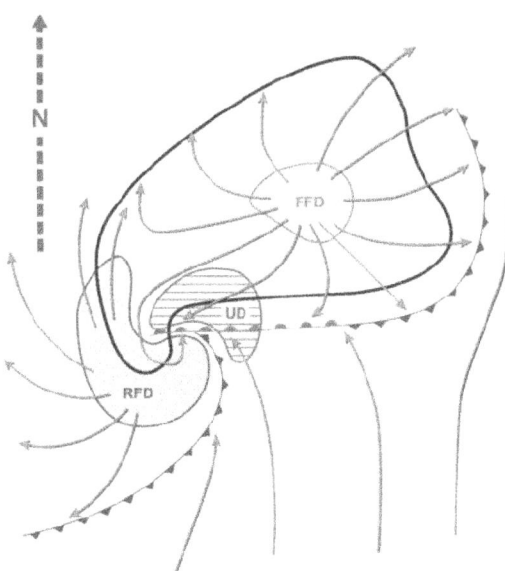

Simplified supercell diagram, rear-flank downdraft (RFD), forward-flank downdraft (FFD), and updraft (UD) areas are highlighted. Arrows show air flow. Blue lines with barbs represent gust fronts. Red line is the edge of the inflow air at ground level. Purple section of line is the occlusion where the surface inflow air gets cut off from the low-level mesocyclone. As a spotter, your location relative to this diagram will influence the winds you experience and the storm features you see.

Supercell Variations

All supercells contain storm-scale rotation, which may give a striated or corkscrew appearance to the storm's updraft. One supercell may appear quite different visually from another, depending on the following:

- The amount of precipitation accompanying the storm
- Whether precipitation falls near or far from the updraft

How a storm moves influences the winds that it "feels," much like sticking your hand out a car window as you turn and change speeds. Winds around the storm play an important role in where precipitation exists in and around the storm's updraft.

Based on their visual appearance, supercells are often labeled as:

- Low precipitation
- Classic
- High precipitation

Low Precipitation Supercell (LP)

- Barber pole or corkscrew appearance is possible.

- Precipitation is sparse or well removed from the updraft below cloud base, which often is transparent.

- Large hail is often difficult to discern visually. Although precipitation may not be apparent below the storm, sometimes very large hail is falling that can not be seen at a distance.

Classic (CL) Supercell

- Majority of supercells in this category

- Large, flat rain-free base

- Can have wall cloud

- Barber pole or corkscrew appearance of updraft possible, as in LP supercell

- Heavy precipitation falls adjacent to the updraft

- Large hail possible

- Potential for strong, long-track tornadoes

High Precipitation (HP) Supercell

- Precipitation often surrounds updraft, and may hide it
- Can have a wall cloud, but it may be obscured by the heavy precipitation
- RFD filled with precipitation
- May have an associated shelf cloud

- Tornadoes potentially obscured by heavy precipitation (rain-wrapped)

- Extremely heavy precipitation with flash flooding

Visual Clues of Supercells

- Rotating wall cloud suggests the presence of a rotating updraft

- Striations on the sides of the storm, streaks of cloud or bands of cloud that give the storm a corkscrew or barber pole appearance, indicate the storm's updraft is rotating, generally seen with Low Precipitation or Classic Supercells

- Inflow cloud bands, such as a "beaver's tail," feed into the storm. The beaver's tail is a smooth, nearly flat cloud band extending out from the eastern edge of the rain-free base toward the east.

Severe Hail and Winds

Hail: NWS issues a severe thunderstorm warning for hail of 1 inch across or larger. When reporting hail, it is best to measure the hail when safe to do so. If you are not equipped with a ruler or other measuring instrument, hail size can be related loosely to coins or athletic balls, as in this table:

BB	Less than 1/4"
Pea	1/4"
Dime	7/10"
Penny	3/4"
Nickel	7/8"
Quarter	1"
Half Dollar	1¼"
Walnut or Ping-Pong Ball	1½"
Golf Ball	1¾"
Lime	2"
Tennis Ball	2½"
Baseball	2¾"
Large Apple	3"
Softball	4"
Grapefruit	4½"

Hail storm. Photo from NOAA.

Damaging Winds: NWS issues warnings when the winds from a thunderstorm are expected to be 58 mph (50 knots) or higher.

Downburst: This term refers to an area of strong, often damaging winds produced by air rapidly descending in a thunderstorm. Downbursts are sometimes described as a microburst when it covers an area of less the 2.5 square miles and last 3-7 minutes, or macroburst when they cover larger areas or last for more than 7 minutes. On rare occasions, downbursts can have wind speeds in excess of 150 mph. Downbursts, both microburst and macroburst, are described as wet or dry.

- Wet: Rainfall accompanies the damaging winds.

- Dry: Very little rainfall accompanies the damaging winds and sometimes all that is visible is the evaporating precipitation (see virga below) or the dust kicked up off of the ground.

Bow Echo: A bow-shaped line of convective cells, best seen on radar, is often associated with swaths of damaging straight line winds and small tornadoes.

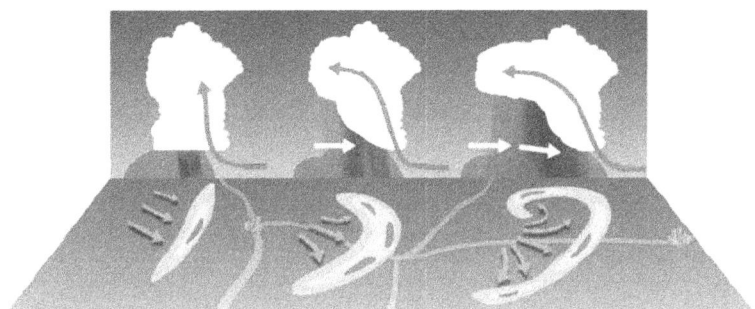

Schematic showing the development of a bow echo.

Derecho (deh-REY-cho): A derecho is a widespread and usually fast-moving convectively induced episode of damaging winds. Derechos can result from bow echoes, supercells, multicell clusters or lines, or a blend of any of these. They can produce damaging straight-line winds over areas hundreds of miles long and more than 100 miles across. Due to the large area extent, an episode is usually not classified as a derecho until it is complete.

Estimating Wind Speed: It's often difficult to estimate wind speed, especially in the plains where there are few physical indicators to observe damage. Below is the Beaufort Wind Force Scale for estimating wind speeds. This is only a rough guide. Actual damage may occur at weaker or stronger speeds.

Wind Speed (mph)	Effects
25-31	Large branches in motion
32-38	Whole trees in motion
39-54	Twigs break off trees, wind impedes walking
55-72	Damage to TV antennas, large branches break off trees
73-112	Surfaces off roofs peeled off, windows broken, trailer homes overturned
113+	Roofs blown from houses, weak buildings and trailer homes destroyed, large trees uprooted, train cars blown off tracks

Town near Desoto, MO, hit by thunderstorms that produced 20 to 30 minutes of severe winds estimated to be between 80 and 100 mph, May 6, 2003. Photo from NOAA.

Spotting Downbursts

There are several visual signs that a downburst is either underway or about to occur.

Virga: Precipitation streaks from the cloud, but does not reach the ground. The atmosphere below the clouds tends to be very dry and rainfall evaporates before it touches the ground. Gusty winds occur in the area of the virga.

Rain foot: The rain foot is a pronounced outward bend of the precipitation area near the ground, marking an area of strong outflow winds.

Flash Floods

- Water rises rapidly with little or no advance warning

- Occurs in steep terrain, but can also occur in urban areas when sewer systems cannot quickly drain the water running off large areas of pavement

- Usually occurs due to heavy rainfall, storm surge, ice jams, dam breaks, or tsunamis

Lightning

- Lightning is an electrical discharge occurring in thunderstorms, but can occur in hurricanes, winter storms, volcanic eruptions and large wildfires.

- Lightning discharges can occur between oppositely-charged parts of the thunderstorm cloud or between opposite charges in the cloud and on the ground.

- In the United States, there are about 20 to 25 million cloud-to-ground lightning flashes each year.

- Lightning kills an average of 30 to 50 people each year in the United States and injures several hundred more.

- Any lightning or thunder indicates a charged atmosphere and consequently a dangerous situation.

- The initial stages of thunderstorm development often bring a short period of in-cloud lightning, heard as a crackling or rumbling aloft. As the thunderstorm continues to develop, you may see cloud-to-ground strikes.

- Most cloud-to-ground lightning is between negative charges in the middle to lower part of the cloud and the positively charged ground. A small percentage of lightning occurs between the positively charged upper part of the cloud and negatively charged ground. These positive flashes can strike over 10 miles from the parent storm.

- As the updraft of a storm weakens, the negatively charged area in the lower part of the storm diminishes and a larger percentage of positive flashes occur.

- The downdrafts from some strong and severe storms pull positive charge downward causing positive flashes near the most intense part of the storm.

- Positive Charged Lightning: This type of lightning makes up less than 5% of all lightning strikes. The majority of positive lightning originates in the upper levels of the storm, and has a much stronger charge to it than most negative-charged lightning strikes. These strikes are very bright, and are known for being good fire starters when they strike grass, brush, or trees. They can strike more than 10 miles away from the parent thunderstorm.

- Negative Charged Lightning: Negative lightning makes up over 95% of all strikes. While the electrical charge isn't usually as strong as a positive strike, it is still enough to kill people who are outdoors without proper shelter.

Tornado Formation

Tornadoes are one of nature's most fearsome creations. Each year they constitute a major hazard around the United States. Storm spotters can help to increase warning lead time by recognizing and reporting clues associated with tornado development and the various stages of the tornado life cycle. This section describes the typically observed features before tornado formation, during the life of the tornado, and as the tornado dissipates.

Visual Clues of Tornado Formation

- Large, rounded rain-free base. This can indicate the presence of a mesocyclone.

- Increasing spin in wall cloud and cloud base around wall cloud. This can suggest that the low level rotation is increasing.

- Clearing skies working into the rain-free base, which suggests a part of the rear-flank downdraft is wrapping around the mesocyclone. This often precedes or accompanies tornado formation in supercell thunderstorms.

- Rapid vertical motions, scud (see p. 49), rising into wall cloud, sinking motion around wall cloud from rear-flank downdraft.

- Local burst of heavy rain/hail just west or southwest of wall cloud. Occasionally, this is a precursor to a tornado.

A tornado may form within a few minutes of these clues appearing! In other cases, outflow behind the gust front can spread out from storm and cut off the formation process. Not all of the signs listed above are required for tornado development; however, storms exhibiting most or all of these features have better odds of producing a tornado.

Developing Stage

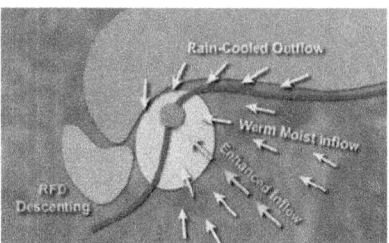

- Tornado circulations can develop from either the ground upward or from the middle and low levels downward.

- Rear-flank downdraft and precipitation southwest of a wall cloud may signal processes that help establish a tornado.

- Some circulations start in low levels, near the cloud base, with rapid accelerations of cloud material into an area of tightening rotation.

- Watch closely! The first sign of tornado development may be a dust whirl at the ground. If seen, closely examine this whirl to see if it is connected to the cloud base.

Mature Stage

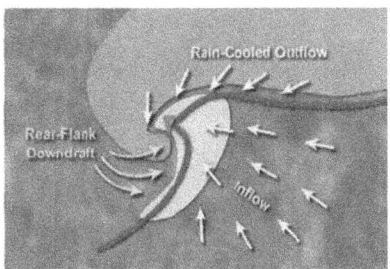

- Potentially the strongest and most dangerous stage of the tornado's lifetime.

- Funnel often has a near-vertical orientation.

- Visible funnel may not extend all the way to the ground, or may become hidden inside the wrapping precipitation!

- Often, rear-flank downdraft wraps around south and east side of the wall cloud gradually cutting off original inflow air.

- Rain-free base may take on a horseshoe-shaped appearance. The tornado and wall cloud may be found at the north end of this structure.

Dissipating Stage

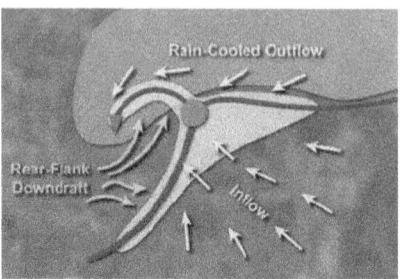

- Rear-flank downdraft wraps around tornado.

- The tornado is separating from the warm buoyant air that it needs for its survival.

- Tornado shrinks, tilts, and takes on a contorted snakelike appearance before finally dissipating. Sometimes this is referred to as the rope stage.

- Although the tornado is not as large as it was in the mature stage, it can still be dangerous.

Cyclic Supercells

- With some supercells, inflow may be refocused a few miles east of the dissipating tornado.

- If the environment is favorable, a new mesocyclone and wall cloud will form.

- The new mesocyclone and wall cloud will become the dominant part of the storm, and a new tornado may form.

- For safety, check overhead often when near the inflow region of a supercell.

Tornado Variations in Appearance

Tornadoes can come in various shapes and sizes. Below are some pictures of the various types that may be encountered.

Multi-Vortex tornadoes have two or more circulations or vortices orbiting about each other or around a common center.

Rope tornadoes often signify a tornado that is weakening or dissipating. Still, such tornadoes can contain deadly and damaging ground circulations, whether visible or not.

Non-Supercell Landspout tornadoes do not arise from organized storm-scale rotation, and therefore usually are not associated with a visible wall cloud or a radar-indicated mesocyclone. Typically, these tornadoes are observed beneath cumulus and towering cumulus clouds, often as no more than a dust whirl, and are the land-based equivalents of non-supercell waterspouts.

A waterspout is a tornado over water. There are two types of waterspouts. The Type A waterspout forms in a supercell thunderstorm and is a violent and potentially destructive vortex that often moves quickly and is capable of significant structural damage if it moves over land.

The Type B waterspout is usually less violent, slower moving, and less destructive. Type B develops quickly beneath a rapidly building line of cumulus clouds. The winds near the base of a Type B waterspout can reach and exceed gale force (34 knots), which is strong enough to swamp or capsize a small watercraft. All waterspouts pose a threat to boater safety and should be avoided.

Tornado/Funnel Cloud Look-a-Likes: Several atmospheric and man-made features may be mistaken for tornadoes. Some of the most common are:

- Scud Clouds
- Rain Shafts
- Gustnadoes
- Tail clouds
- Smoke
- Communication Towers
- Grain Elevators
- Dust Devils

To distinguish between a real tornado or funnel and one of the above look-a-likes, study the feature and be able to answer the following questions:

1. Can I see it clearly?

2. Is the feature attached to a thunderstorm base?

3. Is the feature in the section of the storm where tornadoes/funnels typically develop, i.e., near the updraft?

4. Is there organized rotation present within the feature?

5. If it appears to be a tornado, is there debris?

If your answer to any of these questions is "no," then the feature likely is not a tornado. If you have doubts, continue to observe the feature. Important: report only what you see, not what you think you see.

Scud clouds are low cloud fragments that may attach to a storm's base and can strongly mimic the appearance of a ragged funnel. Some scud can rise from or near the ground leaving the impression of a tornado. Watch for persistent rotation of the suspicious feature to rule out scud.

A Dust Devil is usually a small, rapidly rotating wind that is made visible by the dust, dirt or debris it picks up. Also called a whirlwind, it develops best on clear, dry, hot afternoons.

Technology and Storm Spotting

Doppler Radar

The most effective tool to detect precipitation is radar. Radar, which stands for Radio Detection and Ranging, has been used to detect precipitation, and especially thunderstorms, since the 1940s. Radar enhancements have enabled NWS forecasters to examine storms with more precision.

NWS radars use Doppler weather radar principles. All weather radars, including Doppler, electronically convert reflected radio waves into pictures showing the location and intensity of precipitation; however, Doppler radars also can measure the frequency change in returning radio waves which allows meteorologists to display motions toward or away from the radar.

This ability to detect motion has greatly improved the meteorologist's ability to peer inside thunderstorms and determine if there is rotation in the cloud, often a precursor to the development of tornadoes.

Doppler Weather Radar Images

Reflectivity is the amount of transmitted power returned to the radar receiver after hitting precipitation. It is measured in decibels (dBZ). **Composite Reflectivity** uses all radar elevation scans to create an image and displays the maximum reflectivity vertically at any point. **Precipitation** images (1 Hour and Storm Total) are created by applying computer algorithms to reflectivity imagery to estimate rainfall.

Velocity imagery is a sample of wind data using Doppler principles. Red indicates outbound wind speed and green shows inbound wind speed in knots relative to the radar. **Storm Relative Velocity (SRM)** subtracts the storm motion from the overall wind to reveal winds relative to the storm. This image is useful in displaying small scale circulations within thunderstorms.

Satellite Imagery

NWS satellites are capable of producing information on clouds and moisture in three primary forms: Visible, Infrared (IR) and Water Vapor.

Visible imagery shows the earth in visible light. This process is similar to that of a person taking a picture with a camera. The satellite detects sunlight reflected from objects within the viewfinder. In the case of the satellite, the objects are the upper surfaces of clouds. Thick clouds do a much better job of reflecting light and therefore appear brighter in visible photos.

The problem with visible imagery is that it is only available during the day. To combat this problem, the infrared (IR) sensor was devised. It senses radiant (heat) energy given off by the clouds. Warmer clouds, which are lower in the atmosphere, give off more energy than higher, cold clouds. The IR sensor measures the heat and produces several images based upon different wavelengths in the IR range of the electromagnetic spectrum.

Water vapor imagery is unique in that it can detect water in a gas state in addition to clouds. This type of image shows water vapor in the top one-third of the troposphere. Energy from moisture in the lower levels of the atmosphere is absorbed by the atmosphere and hidden from the satellite sensor. Upper level moist and dry areas are plainly observable and can show prominent air currents. Moist areas show as white, while dry areas show as black.

Types of Satellites

The Geostationary Operational Environmental Satellite's (GOES) path around the earth is at an altitude of 22,236 miles. At this distance the satellite complete one orbit of the earth in 24 hours. The net result is the satellites appears stationary, relative to the earth. This allows it to hover continuously over one position on the surface. Because they stay above a fixed spot on the surface, they provide a constant look at atmospheric severe weather conditions. The United States operates two meteorological satellites in geostationary orbit, one over the equator at 75 deg W with a view of the East Coast and the other over the equator at 135 deg W, with a West Coast view.

Polar Orbiting Satellites (POES) offer the advantage of daily global coverage by making nearly polar orbits roughly 14.1 times daily. Since the number of orbits per day is not an integer, the orbital tracks do not repeat on a daily basis. Currently in orbit are morning and afternoon satellites, which provide global coverage four times daily.

Common Storm Types on Radar

Single Cell Thunderstorm

This type of thunderstorm develops in weak vertical wind shear environments characterized by a single updraft core and a single downdraft that descends into the same area as the updraft. The downdraft and its outflow boundary then cut off the thunderstorm inflow, causing the updraft and the thunderstorm to dissipate. Single cell thunderstorms are short-lived. They only last about a half hour to an hour. These thunderstorms will occasionally become severe (1 inch hail, wind gusts in the excess of 58 mph, or a tornado), but only briefly. In this case, they are called Pulse Severe Thunderstorms.

Multicell Thunderstorm

Multicell thunderstorms are organized in clusters of at least 2-4 short-lived cells. Each cell generates a cold air outflow that combine to form a large gust front. Convergence along the gust front causes new cells to develop every 5 to 15 minutes. The cells move roughly with the mean wind; however, the area (storm) motion usually deviates significantly from the mean wind due to discrete propagation (new cell development) along the gust front. The Multicell nature of the storm is usually apparent on radar with multiple reflectivity cores and maximum tops.

Multicell Line

A line of active thunderstorms, with or without breaks, including contiguous precipitation areas, resulting from the existence of the thunderstorms

Supercell Thunderstorm

Supercell thunderstorms are potentially the most dangerous of the convective storm types. Storms possessing this structure have been observed to generate the vast majority of long-lived strong and violent (F2-F5) tornadoes, as well as downburst damage and large hail. This type of storm consists of one quasi-steady to rotating updraft that may exist for several hours.

Radar will observe one long-lived cell, but small perturbations to the cell structure may be evident. The stronger the updraft, the better the chance the supercell will produce severe weather, hail greater than 1 inch in diameter, wind gusts greater than 58 miles an hour, and possibly a tornado.

Severe Weather Radar Signatures

A hook echo is a radar reflectivity pattern that forms a hook shape, usually in the trailing portion of a Supercell storm. This hook shape forms when precipitation gets wrapped around the storm mesocyclone and is a favorable region for tornado development.

A bow echo occurs when part of a line of thunderstorms accelerates ahead of the rest of the line. This produces a bend, or bow, in the line. This acceleration of the radar echo is a reflection of strong localized "straight-line" winds at or near the surface.

Observations and post-storm analysis show that the greatest threat for straight-line wind damage is typically found near the leading edge, or apex, of the bow.

Another radar characteristic of mature bow echoes is the region of weak reflectivity trailing immediately behind the bowing line of strong thunderstorms. This weakness in the reflectivity is caused by a descending flow of air from mid-levels of the atmosphere.

Sometimes significant wind damage and even weak tornadoes also can occur on the northern end of the bowing line segment within the cyclonically (counter-clockwise) rotating comma head region.

Large, organized, long-lived bow echoes can develop and move across several states producing long swaths of wind damage. This type of convective system is often referred to as a derecho.

In a severe thunderstorm, large water-coated hail stones suspended aloft reflect the radar energy in a complex way. This effects causes a narrow spike of reflectivity to protrude from the intense reflectivity core on the image. This feature is referred to as a three-body scatter spike. The spike is along a radial, the radar beam at that particular azimuth. In basic terms, this is caused by the radar beam hitting the large water-coated hail, scattering the energy to the ground below, then scattering the energy back upward, and finally scattering the energy once again by the hail aloft.

The three scatterings illustrate the triple reflection or Three Body Scatter Spike (TBSS). The presence of a hail spike is a reliable indicator that severe hail, greater than 1 inch in diameter, exists in the storm.

The **bounded weak echo region** (BWER) is a nearly vertical channel of weak reflectivity echoes surrounded on all sides and on top by higher radar reflectivities. The weak reflectivity core is a result of strong storm updrafts carrying hydrometeors upward so quickly they do not grow to radar-detectable size until at high storm levels. The BWER is associated with very strong storm updraft speeds and is typically found 3-10 km above the ground.

The strong updraft speeds associated with the supercells suspend the hail above the updraft until the hail grows too large for the updraft winds to support it. At that time, the hail falls to the ground. The largest hail falls next to the updraft area of the supercell, generally from the west through north side of the mesocyclone.

A **V-notch** is a radar reflectivity pattern that forms a V-shape in the downwind part of a supercell thunderstorm echo. This V-notch is a sign of diverging flow around the very strong storm updraft.

Storm Movement and Spotter Location

Storm spotters must be constantly aware of the storm's location and movement. Since either of these can be hard to determine visually, it is a good idea to take advantage of radar data to help with these critical details. Once you know where the storm is and which way it is moving, you can determine where to position yourself to view the updraft region of the storm. For a storm moving northeast, the best observing location would be to the southeast of the storm. From this direction, you can get a clear view of the rain-free updraft region of the storm where wall clouds and associated tornadoes may form. In any other direction, rain and hail may block the view of the updraft region of the storm.

Never assume all storms move from the southwest to the northeast. Storms typically move in the same direction as the mid-level atmospheric winds, so you may experience storms moving from the west, northwest or even north. Supercell storms sometimes move (turn) to the right of the mid-level winds, and these storms typically have a higher potential of to become severe. Remember storms don't move in a straight line at a constant speed. It is critical to have current information about the motion and behavior of storms in your area and to use this information to avoid the most dangerous parts of the storm.

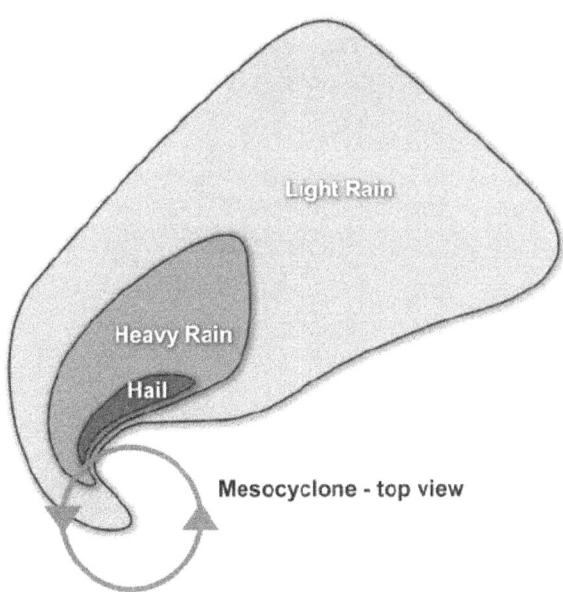

Mesocyclone - top view

.

www.ingramcontent.com/pod-product-compliance
Lightning Source LLC
Chambersburg PA
CBHW070227210526
45169CB00023B/1016